Essential Jobs

We Need Energy Workers

by Brienna Rossiter

FOCUS READERS
PIONEER

www.focusreaders.com

Focus Readers is distributed by North Star Editions:
sales@northstareditions.com | 888-417-0195

Produced for Focus Readers by Red Line Editorial.

Photographs ©: Shutterstock Images, cover, 1, 4, 7, 8, 11, 12, 15 (top), 15 (bottom), 17, 18, 21

Library of Congress Cataloging-in-Publication Data
Library of Congress Cataloging-in-Publication Data is available on the Library of Congress website.

ISBN
978-1-63739-030-6 (hardcover)
978-1-63739-084-9 (paperback)
978-1-63739-191-4 (ebook pdf)
978-1-63739-138-9 (hosted ebook)

Printed in the United States of America
Mankato, MN
012022

About the Author

Brienna Rossiter is a writer and editor who lives in Minnesota.

Table of Contents

Chapter 1

Energy

People use energy in many ways. People use energy to heat homes. They use it to drive cars. Lights and **appliances** use energy, too. All this energy comes from many sources.

Many workers help produce energy. Some find or make **fuels**. Others get fuels ready for people to use. Workers also study how using energy **affects** the planet.

Fun Fact

Some energy sources are **renewable**. Others can run out.

Chapter 2

Fossil Fuels

Some energy comes from **fossil fuels**. Coal is one example. Oil is another. So is natural gas. These fuels are buried underground. Workers dig or drill to find them.

Next, workers get the fuels ready to use. They do this at factories. They help clean or **refine** the fuels.

Then, workers send the fuels out. They may use pipes. They may use ships or trains.

Chapter 3

Renewable Energy

Some energy sources are renewable. For example, people get energy from wind. They can use sunlight, too. People use machines to make this kind of energy.

Some workers build machines. Some fix them. Others find ways to **improve** the machines. They plan how to make more power. Workers also find new ways to make fuel.

Fun Fact

Dams use moving water to produce power.

On the Job

Nuclear Power

Some places use nuclear power. This power is made by splitting **atoms**. This process makes lots of power. But it can be risky. It uses **unstable** fuel. It also makes dangerous waste. So, nuclear power plants have many workers who help with safety. Some watch for problems. Others help store the waste.

Chapter 4

Electricity

Energy is often turned into electricity. Power plants help do this. Many people work at power plants. They check and run machines.

Wires bring electricity to homes and buildings. Some workers make or fix these wires. Other workers watch for problems. If the power goes out, workers find out why. Then they help fix the problem.

FOCUS ON

Energy Workers

Write your answers on a separate piece of paper.

1. Write a sentence describing one way that people use energy.
2. Which type of fuel do you think is most important for workers to make? Why?
3. Which type of fuel are coal, oil, and natural gas?
   A. fossil fuels
   B. renewable energy
   C. nuclear power
4. Windmills use wind to make electricity. Which type of energy do they produce?
   A. fossil fuels
   B. renewable energy
   C. nuclear power

Answer key on page 24.

Glossary

affects
Changes or shapes something.

appliances
Machines made to do certain jobs. Stoves, toasters, and fans are some examples.

atoms
Very tiny bits of matter.

fossil fuels
Energy sources that come from plants and animals that died long ago. Fossil fuels include coal, oil, and natural gas.

fuels
Materials that are burned to create heat or power.

improve
To make something better.

refine
To remove unwanted parts or pieces.

renewable
Possible to get or make more of.

unstable
Likely to move, change, or fall apart.

To Learn More

BOOKS

Felix, Rebecca. *Wind Energy*. Minneapolis: Abdo Publishing, 2019.

Zalewski, Aubrey. *Renewable Energy vs. Nonrenewable Energy*. Mankato, MN: The Child's World, 2020.

NOTE TO EDUCATORS

Visit **www.focusreaders.com** to find lesson plans, activities, links, and other resources related to this title.

Index

Answer Key: 1. Answers will vary; **2.** Answers will vary; **3.** A; **4.** B